Special Report:
Arson and Juveniles:
Responding to the Violence

A Review of Teen Firesetting and Interventions

Paul Schwartzman
Hollis Stambaugh
John Kimball

This is Report 095 of the Major Fires Investigation Project conducted by Varley-Campbell and Associates, Inc./TriData Corporation under contract EMW-94-C-4423 to the United States Fire Administration, Federal Emergency Management Agency.

 **Homeland Security**

Department of Homeland Security
United States Fire Administration
National Fire Data Center

U.S. Fire Administration Fire Investigations Program

The U.S. Fire Administration develops reports on selected major fires throughout the country. The fires usually involve multiple deaths or a large loss of property. But the primary criterion for deciding to do a report is whether it will result in significant "lessons learned." In some cases these lessons bring to light new knowledge about fire--the effect of building construction or contents, human behavior in fire, etc. In other cases, the lessons are not new but are serious enough to highlight once again, with yet another fire tragedy report. In some cases, special reports are developed to discuss events, drills, or new technologies which are of interest to the fire service.

The reports are sent to fire magazines and are distributed at National and Regional fire meetings. The International Association of Fire Chiefs assists the USFA in disseminating the findings throughout the fire service. On a continuing basis the reports are available on request from the USFA; announcements of their availability are published widely in fire journals and newsletters.

This body of work provides detailed information on the nature of the fire problem for policymakers who must decide on allocations of resources between fire and other pressing problems, and within the fire service to improve codes and code enforcement, training, public fire education, building technology, and other related areas.

The Fire Administration, which has no regulatory authority, sends an experienced fire investigator into a community after a major incident only after having conferred with the local fire authorities to insure that the assistance and presence of the USFA would be supportive and would in no way interfere with any review of the incident they are themselves conducting. The intent is not to arrive during the event or even immediately after, but rather after the dust settles, so that a complete and objective review of all the important aspects of the incident can be made. Local authorities review the USFA's report while it is in draft. The USFA investigator or team is available to local authorities should they wish to request technical assistance for their own investigation.

For additional copies of this report write to the U.S. Fire Administration, 16825 South Seton Avenue, Emmitsburg, Maryland 21727. The report is available on the Administration's Web site at http://www.usfa.dhs.gov/

U.S. Fire Administration

Mission Statement

As an entity of the Department of Homeland Security, the mission of the USFA is to reduce life and economic losses due to fire and related emergencies, through leadership, advocacy, coordination, and support. We serve the Nation independently, in coordination with other Federal agencies, and in partnership with fire protection and emergency service communities. With a commitment to excellence, we provide public education, training, technology, and data initiatives.

ACKNOWLEDGMENTS

The United States Fire Administration appreciates the help of the following organizations that provided information for this report:

Cobb County Juvenile Court Mediation Program, Marietta, Georgia

Lane County Department of Youth Services, Springfield, Oregon

Marietta Fire Department, Marietta, Georgia

Oregon State Fire Marshal's Office, Salem, Oregon

Phoenix Fire Department, Phoenix, Arizona

Portland Fire Department, Portland, Maine

Portland Fire Department, Portland, Oregon

Providence Fire Department, Providence, Rhode Island

Rochester Fire Department, Rochester, New York

TABLE OF CONTENTS

INTRODUCTION

The purpose of this U.S. Fire Administration special report is to document the problem of older children who set fires resulting in serious or potentially serious consequences. The report also examines the factors that commonly are associated with **intentional** firesetting by teenagers and discusses a number of community programs that intervene to control arson.

Historically, the term "juvenile firesetting" has been viewed as a "curious" kids' problem. Fires set by youngsters playing with matches and lighters tend to be categorized as "accidental" or "children playing." However, juvenile firesetting also includes the deliberate destruction of property by juveniles through fire, which sometimes results in casualties. This is an increasingly serious problem in most U.S. cities. Information from a 10-year U.S. Fire Administration project of direct technical assistance to over 60 jurisdictions verifies the high rate of juvenile-set fires.

This report focuses on a adolescent firesetters between 14 and 18 years of age. Several case studies are presented to demonstrate the impact of these arson fires and to outline the family circumstances of the youth who were involved. The report also covers how the criminal justice system has been handling teen arson and reviews and compares several treatment and intervention programs.

OVERVIEW OF THE PROBLEM

Arson is the number one cause of all fires (approximately 550,000 in 1994), and the second leading cause of residential fire deaths. Five hundred sixty-five fire deaths and 3,440 fire injuries in 1994 were attributed to arson.

The dollar loss from arson fires was estimated at $3.6 billion for 1994. According to insurance industry reports, the average property loss from incendiary and suspicious fires in 1996 increased by 24 percent from the year 1995 to $27,810. The loss of valuable properties reduces the property tax revenues necessary to support public safety agencies including municipal fire departments.

Fire service data compiled by the National Fire Incident Reporting System (NFIRS) have repeatedly shown that firefighter injuries are significantly higher at arson fires than at accidental fires. Arson fires account for 22 percent of firefighter injuries.

The Federal Bureau of Investigation (FBI) Uniform Crime Report from 1995, the most recent year that complete data is available, indicates that juveniles account for the majority of arson arrests. **Fifty-two percent of arson arrests include children under the age of eighteen.** While national indicators of juvenile violent crime are suggesting that incidents such as murder and aggravated assault are on the decline, **the incidence of juvenile arson continues to increase**. In the early part of the 1990's, juvenile arson arrests remained constant at about 40 percent. In 1993, the figure was 49 percent. The majority of those arrested for arson in 1994 were under 15, and nearly 7 percent were younger than 10.

Using FBI statistics and National Incident Reporting System data, it is estimated that there are at least 100,000 fires annually in the United States directly attributable to children. It is widely believed that this number is conservative due to the fact that many fires never come to the attention of the fire service. According to the ninth edition of "Fire in the United States," the ratio of reported fires to unreported fires is about three to one. In many states, statutes do not allow younger children to be

1

charged with arson, and many are reluctant to label a child as an arsonist. In fact, if the percentage of juvenile arrests is applied to the total number of incendiary and suspicious fires that occurred in 1993, there were potentially 250,000 fires attributed to juveniles.

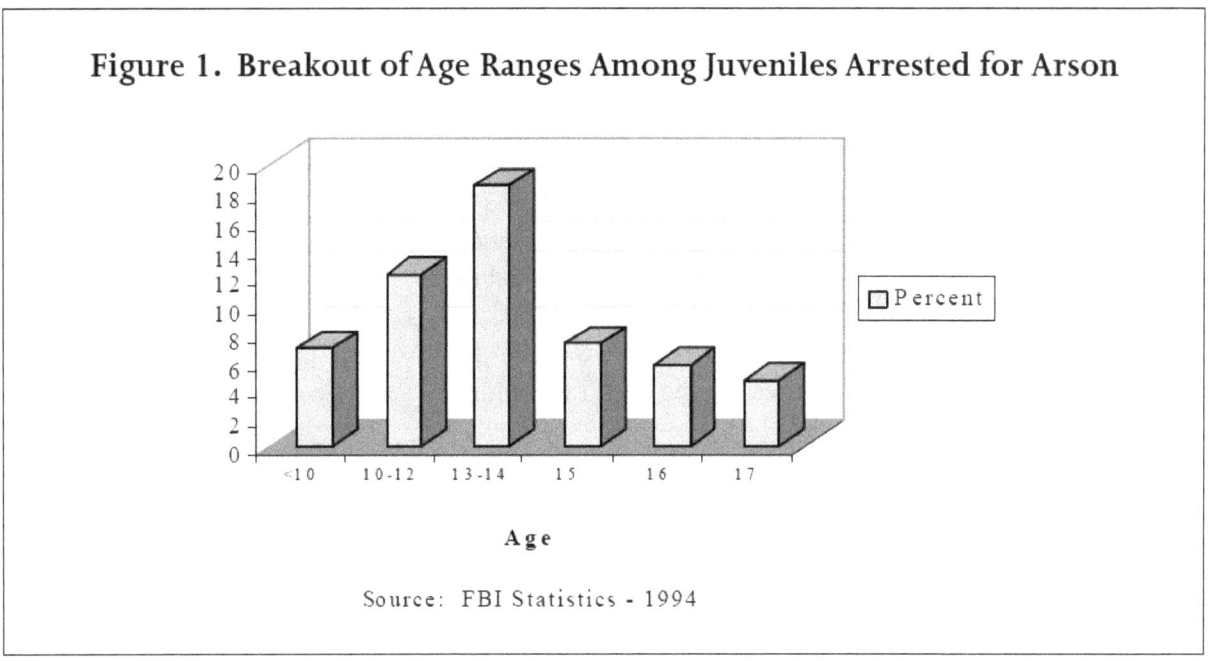

Figure 1. Breakout of Age Ranges Among Juveniles Arrested for Arson

Source: FBI Statistics - 1994

In 1994, two-thirds of all arson fires occurred outdoors. Previous research suggests that as children get older, their firesetting tends to be directed away from their own homes and involves locations such as garbage dumpsters, barns, vacant buildings, grasslands, automobiles, and schools. According to an aggregation of statistics from USFA, the National Fire Protection Association (NFPA), and the Federal Bureau of Investigation, twenty percent of arson fires occurred in structures, thirteen percent in automobiles, and sixty-six percent in the outdoors, primarily trash and grass fires. These percentages have remained fairly constant for more than a decade.

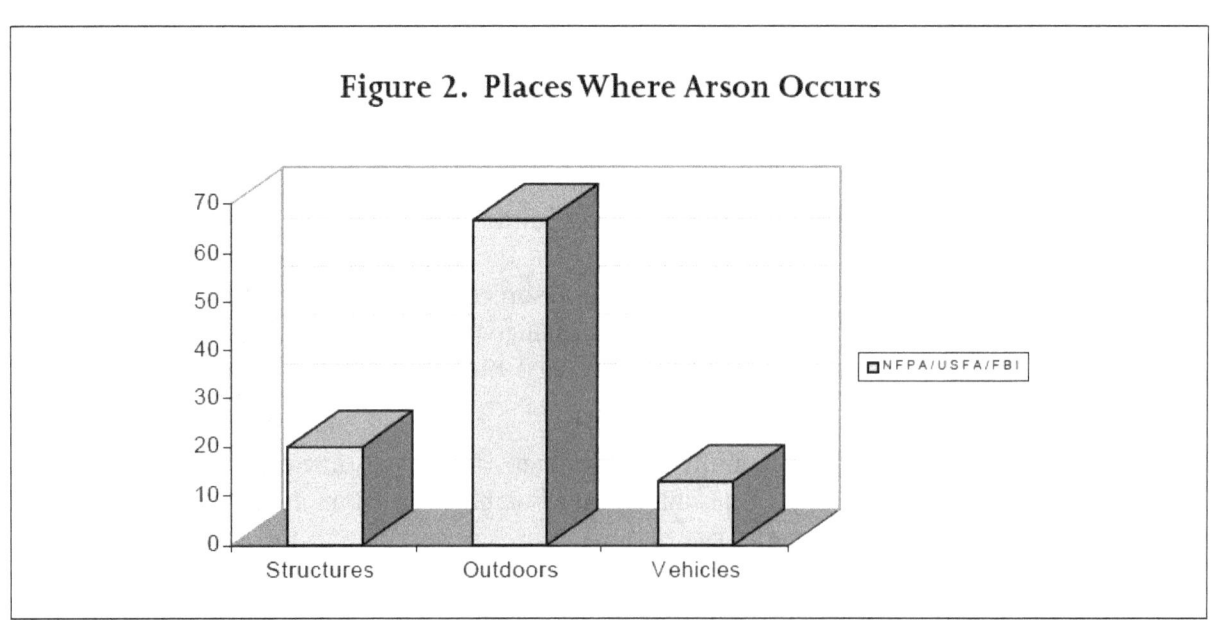

Figure 2. Places Where Arson Occurs

As part of the research for this report, a sampling of 35 incendiary fire cases involving older juveniles was reviewed. The cases were provided by eight selected fire departments. Figure 3 provides a break out of where the juveniles started fires. **Twenty-six percent of these fires took place in occupied dwellings or schools,** 37 percent took place in abandoned houses and buildings, and 37 percent occurred outdoors in dumpsters, parks, or open areas. This 35 case sample was not necessarily representative due to the fact that cases were hand selected against specific criteria: a serious consequence to the firesetting, which builds in a bias toward occupied structures.

Figure 3. Study Cases of Serious Incendiary Fires Set By Older Juveniles

Property	Number	Percentage
Occupied dwellings and schools	9	26
Abandoned structures	13	37
Dumpsters, park, open areas	13	37

Although it appears that unoccupied buildings and outdoor areas (especially where there is debris or dumpsters) are at greatest risk for juvenile arson, a significant amount of fires are set in occupied structures, indicating that intentionally-set juvenile fires can have very serious consequences. When motive is factored in, a pattern tends to emerge both in the study cases and from local experience dealing with juvenile-set fires. Juveniles who set fires to bring attention to difficult family circumstances are more likely to target occupied structures like their homes or schools. Gang-related and revenge fires on the other hand, occur more often in abandoned buildings (often used as drug houses or places to meet), but rarely in the offender's own home.

JUVENILE-SET FIRES COST LIVES AND PROPERTY

Omaha, NE; April 25, 1996

A Captain in the Omaha Fire Department was killed while fighting a fire set by a 15-year-old boy. The blaze was located in a department store and the fire captain was trapped when the roof collapsed on him. The cause of the officer's death was smoke inhalation. Omaha police considered the case a homicide.

Philadelphia, PA, September 19, 1994

Three youths, ages 13, 15, and 16 were hired by local drug dealers to set a fire in a vacant factory. The facility, the former Quaker Lace plant, was 5 stories high and covered most of a city block. Salvage and removal of heavy manufacturing equipment from the building was underway. While the structure was in the process of transition, one corner of it was being used by the police to monitor drug traffic in the neighborhood. A group of local drug dealers recruited the boys to burn the area used for observation.

The ensuing fire destroyed the entire factory and spread to the neighborhood, forcing the evacuation of 47 families. In all, 20 occupied properties and 11 automobiles were destroyed in the fire. The impact of the incident was so great that it provided the impetus for establishment of the Eastern Philadelphia Drug and Arson Task Force (EPDART), which remains in existence today.

Philadelphia, PA, March 1996

In a different neighborhood, seven youths under the age of 18 began setting fires in dumpsters, then graduated to automobiles and vacant buildings. In March of 1996, members of this group set a fire in an illegal tire dump beneath Interstate 95 resulting in damages estimated at $8 million to the overpasses. Commuter and interstate traffic was disrupted during the incident and for months during the repair. Various members of this group are linked to intentionally set fires in 18 vacant buildings in the same area.

Earlington, KY, April 5, 1997

Two teenage boys were charged with murder and arson when the fire they set in a three-story apartment building trapped and killed three people.

Aloha, OR, June 28, 1996

A twelve-year-old was determined to be criminally responsible for the deaths of eight people including five children aged 3 months to 10 years. He set the fire in an apartment stairwell using newspaper and rubbing alcohol. He was reportedly abused by his father as an infant and was subject to an alcohol fire set by the father, who is in prison on robbery charges.

FIRE DEPARTMENT RESPONSE TO SUSPECTED JUVENILE INVOLVEMENT

Juvenile firesetting is a community problem, and the fire service is in a unique position to address it. The fire department has the job of detecting the problem, investigating the fires, and initiating a response ranging from educational intervention to prosecution. The fire service should make certain that the appropriate referral or action is taken. Documentation of observations made by company level fire crews and officers can be a critical link in the chain of arson recognition and intervention.

Generally, it is not difficult to ascertain juvenile involvement in set fires. Often, the characteristics of the fires present strong indicators that juveniles committed the crime. Discussions with fire investigators and a review of arson incident reports suggest several factors that are critical when solving juvenile-set fires. Many of these points relate to adult-caused incendiary fires as well.

- All fires set by juveniles need to be taken seriously. The size of the fire and the amount of damage are not good indicators of risk. Very often, juveniles who set fires start with small insignificant fires, then graduate to bigger, more daring blazes as they acquire confidence and experience. Fire investigators should address today's small fires as though they could become tomorrow's fatal, multiple alarm fires.

- An immediate and systematic response is essential. As with other fires, investigators should respond to the scene and interview first arriving firefighters and available witnesses. Collecting witness information is one of the most critical parts of fire investigation. If investigations are delayed, witnesses can be difficult to track down. Even if they can be located, witnesses often are more hesitant to cooperate and provide less useful information after leaving the scene.

- Careful observation of the people watching the fire can help identify a firesetter. One investigator noticed that a certain young man tended to be present at every vacant building fire in a particular neighborhood during a three-month period. The youth was also anxious to talk to firefighters about the fires.

- As is true in all incendiary and suspicious fires, preservation of evidence and thorough origin and cause investigations are very important. When questioning adolescents, especially resistant adolescents, knowing several facts in advance about the fire can help determine the truth. An investigator was attempting to determine the cause of a bedroom fire that resulted in several thousand dollars damage to a single-family home. After looking at the fire damage to furniture and other articles in the room, he determined that the point of origin was the north wall which had a baseboard heater and burned debris in front of it. A careful review of the heater and thermostat showed no signs that it had malfunctioned or overheated. The firesetter apparently had stuffed clothing and boxes into the heater, thus precipitating the fire.

  The adolescent originally reported that he entered the room, smelled smoke, heard the smoke detector, and called the fire department to report "flames from the baseboard heater." When confronted with the physical evidence of the case, the teenager amended his story and confessed to the truth.

Other cases have been solved by systematic interviews with school personnel, neighbors, other adolescents, and personnel from other agencies, such as recreation leaders, who have contact with youth and may overhear their stories of conquest and accomplishment. Very often, teens brag about their deeds to one another. Teachers can provide information about conflicts and about disgruntled students. It was a teacher's information which helped solve the multi-million dollar school fire described in the first case study.

IDENTIFYING OLDER JUVENILE FIRESETTERS

Like younger children who are involved in fire play and firesetting behavior, their older counterparts are not a homogeneous group; they come from a variety of family circumstances.

Age

Examining the 14- to 18-year-old group more closely, records from several fire departments show that the vast majority are in the lower end of that age range. One participating department reviewed all their incidents for the past four years. They reported 876 referrals to the juvenile unit between July 1, 1994, and June 30, 1997. Fourteen- to eighteen-year-olds made up 97 (11 percent) of the referrals during that period. When this group was screened for fires with serious consequences or strong potential of serious consequences, thirteen children were identified, or 1.5 percent. Of these, eight were fourteen-year-olds, four were fifteen-year-olds, and one was sixteen years old. There were no 17- or 18-year-olds categorized in this sample. This is consistent with FBI arrests statistics where the majority of juvenile arson arrests were youths 15 years of age and under. Incidents from other fire departments in this study echoed these results.

Figure 4 illustrates the age breakdown of juvenile arson arrests statewide in Oregon. No national data other than FBI arrests currently exists with respect to age categories. NFIRS is scheduled to begin collecting age data which will allow more comprehensive study of distribution by age of arrested party.

Figure 4. 1995 Oregon Juvenile Arson Arrest Data

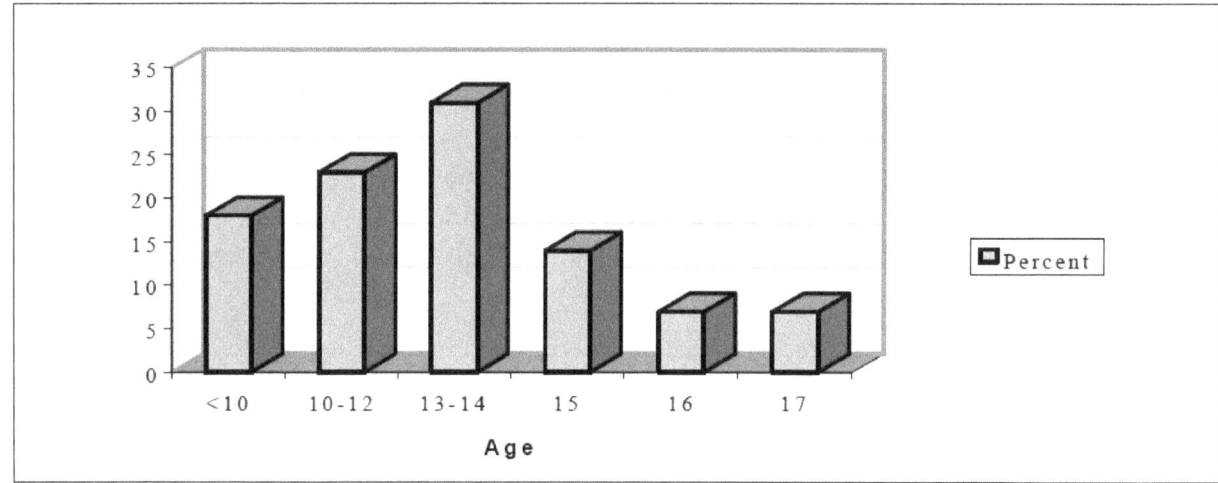

The majority of children in the cases studied lived with their biological parents in an intact family situation or with a biological parent in a single-parent situation. A few children lived in step families or with other relatives; one young woman was homeless. Studies conducted in Rochester, New York, document that type of family unit is not a predictor of recidivism in firesetting. **A large number of family problems are the strongest predictors of recidivism.**

Child Characteristics

Children start fires for varying reasons. Although curiosity is a primary motivation for younger children, it should be noted that curiosity can also be a reason why older teens set fires. Adolescents are attempting to assimilate into an adult world and manipulate adult tools to learn and acquire a sense of control. Often their experimentation is more sophisticated than the experimentation of younger children. Unfortunately, it may involve higher risk substances, such as fireworks or flammable liquids.

Clinical studies that have examined juvenile firesetters find that many of these children have a plethora of conduct and aggression problems. Some children are diagnosed as having attention deficit-hyperactivity disorder.

In a sample of hospitalized firesetters, Dr. David Kolko at the University of Pittsburgh Medical Center found greater delinquency, aggressiveness, and hyperactivity compared to a control group of hospitalized children with no history of firesetting. He also documented that these children were less socially skilled and more aggressive. The children in the case illustrations support these findings. Many of the children also presented with learning disabilities; one was diagnosed with fetal alcohol syndrome, and one suffered a head injury which resulted in a change of behavior and a seizure disorder.

FAMILY PROBLEMS AND RECIDIVISM

Many of juveniles who turn to firesetting are exposed to problems ranging from poor parental judgment and parenting skills to chronic neglect and abuse. Parental alcohol and substance abuse is also not uncommon. In some cases reviewed, a parent suffered from a chronic illness which resulted in unemployment, poverty, and major family problems.

Clinical studies (Cole et al., 1983, 1986; Kolko & Kazdin, 1985, 1986, 1990, 1991) show significant evidence for parental and family problems in families of firesetters. These studies describe parents as significantly lower in affection, depressed, unavailable, and lacking in supervision and parenting skills. The Rochester studies (Cole et al., 1983, 1986) strongly document a correlation between abuse, chronic neglect, and firesetting. In families where there was a founded case of abuse and/or neglect, there was a fivefold increase in the likelihood of recidivism. Also, strongly related to recidivism was a family history of police contact--further indicators of family chaos.

Figure 5. Predicator of Recidivism Rochester, NY, Incident Sample

Prior Family Contact	Number of Non-recidivists	Number of Recidivists	Percent Recidivism
Police			
No	457	11	2.4
Yes	63	6	8.7
Total	520	17	3.2
Child Protective Services			
No	467	11	2.3
Yes	70	9	11.4
Total	537	20	3.6

Several studies have suggested that children who set fires are under stress (Bumpass et al., 1983; Cole et al, 1983, 1986; Fineman, 1980, Health et al., 1983; Jacobson, 1985). They are responding to major life changes such as separation and divorce, remarriage, or death of a family member. This is certainly illustrated in a number of the case studies collected for this project.

MOTIVES

Juvenile firesetting has been studied for several years and there is a general consensus as to what motivates children to become involved with fire. Curiosity motivates a significant portion of fire involvement. Developmental studies report that 40 percent of all children have engaged in fire play. These children are by nature risk takers and learn by doing. This trait combined with ready access to matches and lighters, the belief that parents would not punish them, a poor understanding of fire, and lapses in supervision, accounts for many thousands of fires every year.

The children studied in this project were primarily motivated by something other than curiosity and their firesetting was intentional. There are three basic categories with the first being children whose firesetting is a call for help or attention. Many of these children live in difficult circumstances and lack support. Further, they do not possess appropriate or effective communication skills; their personal observations and experiences have reinforced feelings of isolation and rejection.

The City of Rochester (NY) School District has been conducting a pilot project for the past few years called the Long Term Suspension Project (LTS). The goal of the project is to reconnect students who have been suspended from school for a long period of time and help them to be successful upon their return to school. Long-term suspensions from school are enforced for weapons possession; assault; possession of alcohol and/or drugs; and firesetting. The initial task of the LTS project was to acquire an in-depth understanding of these adolescents and their families. The initial phase of the project focused on high school age youth with later phases incorporating middle school students.

One of the findings of the LTS project was that many of these youths learn the "power of fire" quite early. Their firesetting experience may have started with curious experimentation, but they soon realized that fire got a reaction from parents, authorities, and emergency services. It is a powerful means of communication that is literally at their fingertips.

Another cause of firesetting among the study youth involved delinquent activity, usually carried out in groups in response to peer pressure and/or gang activity. While many of the dynamics involved in attention-seeking behavior are relevant to this group, very often the act of starting a fire is arbitrary. If incendiary materials are handy, they start a fire, and if a rock is handy, they throw it through a window. The motive may involve revenge. In some communities, territorial disputes between gangs over drug trafficking encourage firesetting behavior. The use of molotov cocktails is increasingly prevalent, as has been reported in a number of these incidents.

The Phoenix Fire Department has labeled juvenile gangs as "strategic firesetters". Phoenix has seen an increase in these types of fires and, with the help of Dr. Jeffrey Thomas, has closely examined the dynamics surrounding their behavior. Dr. Thomas describes strategic firesetters as teenagers who have a history of involvement with the juvenile justice system and/or mental health system. Most have been unsuccessful in school. They may have a history of alcohol and substance abuse. These strategic firesetters demonstrate behaviors indicating poor self-esteem and little regard for human life. As a result, they do not show guilt or remorse for sociopathic behavior, including violence against people and property.

Strategic firesetters generally set fires as a group. Fire investigators report that these fires typically involve the use of accelerants and often have multiple points of origin. The fires are set for the purpose of revenge, to instill fear in a community, or to destroy evidence from another crime. When confronted, the strategic firesetter is usually resistant and uncooperative.

The firesetting motivation for another set of older intentional firesetters relates to severe emotional disturbance. Very often, these are children who have been exposed to chronic family dysfunction and situational abuse for long periods of time without sufficient intervention. Their conduct disorder has become quite internalized and is relied upon as a coping mechanism. Some children are motivated out of emerging psychoses or other serious mental illness.

Among juveniles, arson for hire is uncommon, but not unheard of. Recently, a 44-year-old landlord in New York City was arrested after paying a teenager $4,000 to drug the tenants with heroin, and then set fire to their apartment so the landlord could get rid of the occupants and charge higher rent. The case of the Quaker Lace fire in Philadelphia, which destroyed 20 properties and 11 cars, shows the impact of these "arson-for-hire" fires.

CASE EXAMPLES

Case #1: 15-year-old male

Early one evening, a boy broke into his school with the intent of burning it. He started three separate fires in different locations to ensure that his effort would be successful. He left the school and waited. Nothing happened. Frustrated, he returned to the school, broke in a second time, and reignited the fires. This time his effort resulted in a multiple alarm fire which caused $3.5 million damage to the school building.

The boy lives in an upper-middle class neighborhood in a stable home environment. He lives with his biological mother and stepfather. His biological father is not really involved in his life, but all indications were that this was not an issue to him. No other significant family stressors were reported. However, it was indicated that his parents had poor parenting skills and judgment and would often allow him to come and go as he pleased. This lack of structure and clear expectations led to persistent school problems which resulted in his being reprimanded in school the day of the fire. The boy stated he was angry at his teachers and wanted to burn the school down.

Case #2: 15-year-old male

A teenage boy lived in an abandoned trailer with his mother. His father had deserted him years before. His mother was a drug addict who often disappeared for periods of time, leaving him completely alone with no support or means to care for himself.

In his frustration and anger at his mother's absence, he set nine fires in one night. The fires were all started near occupied structures. One was ignited on a front stoop. Several were in dumpsters near residences. Although the potential for loss was significant, none of the fires resulted in major damage.

Case #3: 14-year-old male

A teenage boy was out early one morning with a few of his friends. He was proud to say he was a member of a street gang and had shot at people in the past. He and his friends decided to steal two cars and go for a ride, picking up some additional friends along the way.

While stealing the first car, the boy started a fire in a garage attached to a single-family dwelling. There were paper sacks on the floor next to the car, and he impulsively ignited the material using a lighter and a spray perfume bottle to simulate a torch. He stated that the fire appeared to be going out when they left the garage in the stolen car. However, the fire flared up and spread to the exterior of the house, causing several thousand dollars damage.

He said he did not know why he lit the fire. He and his friends were apprehended after they crashed the stolen cars.

Case #4: 16-year-old female

An arson fire occurred in a vacant single-family dwelling one evening around 9:00 p.m. Alerted to the fire, neighbors ran to the home to discover a teenage girl in the house. Neighbors who urged her to leave the house stated they heard her say, "I started the fire, isn't it pretty?"

The home had been unoccupied since the death of its former resident. However, the police received numerous reports of vagrants and of drug-related activities. The 16-year-old girl explained that she had moved out of her parent's home to the streets exactly one year ago, and that she had stayed in the house on about 15 occasions. The night of the fire she entered through an open back door and started a fire in the fireplace, using papers for heat. Some papers fell out of the fireplace onto the floor. She attempted to fuel the fire rather than extinguish it. Intending to burn down the house, she also started fires in four more locations. When asked why she didn't leave, she stated that her mind was in the gutter and she couldn't think straight. The girl was under the influence of drugs at the time. She denied any suicidal intent and was placed in detention and referred for evaluation.

Case #5: 15-year-old female

A girl was expelled from school after she and a friend singed the hair of two other girls by using hair spray and a lighter to make a torch. The teenager frequently was in trouble at school. The investigator was very concerned about her lack of empathy and remorse for her violence against the two girls. The father stated he believed that his daughter was aware of what she was doing, and that she wanted to cause harm. He is frustrated and tries to monitor her behavior. She was referred for further evaluation.

Case #6: 15-year-old male

A boy admitted starting a fire by putting plastic bags, clothing, and boxes in a baseboard heater in a spare bedroom of his home. The resulting fire caused $60,000 damage to their single-family home.

The boy had a history of fire play and had been referred to the local juvenile firesetter program three years before. At that time, he had started a fire in a closet because he wanted to be a firefighter. Later, the boy admitted to willingly causing the fire. His father had a chronic illness and it appeared that the boy had to manage household responsibilities that he resented. He did not feel that he was properly acknowledged for his increased responsibility. When asked about the incident, he stated that he was angry at his parents.

Case #7: 15-year-old male

After weekly occurrences of dumpster fires behind the local school, the arson investigation unit established a stake out. In full view of an investigator, a teenage boy started a fire in the dumpster, and was quickly apprehended. When asked to explain his actions, he stated that he liked the excitement of doing something bad and getting away with it. He found fire to be especially interesting. He had started fires in dumpsters every Sunday morning for several weeks.

This teenage boy admitted to several other fire incidents dating back to when he was seven years old. Almost all of his firesetting occurred when he was alone. Other intentional fire incidents included a grass fire, setting his school desk on fire, dismantling fireworks and making homemade fireworks, and burning paint thinner in the kitchen sink. Recently, he sprayed lubricant into a glass bottle, then held a lighter to the end of the bottle. The vapor ignited and burned his thumb.

The boy lived with his mother and father. Both parents were unemployed, his father had chronic health problems, and his mother was an alcoholic.

Case #8: 18-year-old male

This older teen left his 12-year-old girlfriend and her mother late one afternoon, promising to return with some fast food. Having no money he decided to break into a cold storage warehouse facility searching for something of value to fence. Once inside, he ignited some large paper bales for no apparent reason. He left the plant without finding anything of value. The ensuing fire required more than 200 firefighters and 50 pieces of fire apparatus to bring under control. The fire destroyed several connected businesses with a loss of about 25 jobs and an insurance claim exceeding two million dollars. This was the largest incendiary fire in Massachusetts in 1995.

Case #9: 14-year-old male

The boy, along with another 14-year-old boy and a 15-year-old girl, broke into a huge idle mill. For several days they explored the premises, performing numerous acts of vandalism. On the last day they began to set small fires on each floor. One of the fires kindled a wood wall and extended into the ceiling. The resultant general alarm arson fire destroyed the plant and endangered a large number of surrounding, occupied multi-family residences.

IMPACT OF TREATMENT PROGRAMS ON RECIDIVISM

There is no doubt that in the United States there is a crisis involving youth aggression and violence. Pressure is being placed on the juvenile justice system to respond to this problem, and to enlist resources from the community. The juvenile justice system has a critical role. It has the power to mandate services and to hold juveniles and their families accountable. Given the chaos in the lives of most older juvenile firesetters, and the documented risk to the community, this power is a pivotal factor in controlling juvenile crime, including arson.

There is a movement in this country to treat serious teen offenders as adult offenders with the belief that more severe punishment will deter this behavior. One recent study conducted at the University of Central Florida by Donna M. Bishop and her colleagues examined more than 2,000 juvenile offenders who were transferred to adult courts. These juveniles were compared with a carefully matched sample of offenders who were retained in the juvenile corrections system. Recidivism was examined in terms of rates of reoffending, seriousness of reoffending, and time to failure, with adjustments made for time at risk. By every measure, reoffending was greater among juveniles transferred to the adult criminal system.

The juveniles in the study who were transferred to adult courts were treated more harshly and, typically, were incarcerated for longer periods of time. Despite this, the transferred youths committed more offenses after they were released, and these offenses tended to be felonies. When comparisons were made between pre and post levels of offending among those who reoffended, the youths retained in the juvenile system generally improved their behavior over time. When they were rearrested, it tended to be for lesser offenses.

Several states have instituted "shock incarceration" or boot camps in an effort to reduce recidivism in all categories of crime. These are rigorous programs in which offenders participate in military-style boot camps, which emphasize drills, physical training, and hard labor, lasting from three to six months. The programs that demonstrated success in reducing recidivism, however, included intensive supervision and follow up for six months following incarceration, with a strong focus on rehabilitation and skill building.

A recent study of juvenile boot camps (Peters, Thomas, Zamberlan, 1997) in Ohio, Colorado, and Alabama reported similar results when looking at recidivism. However, adolescents placed in the camps usually improved their academic performance at the completion of their stay. The study emphasized that there are many differences in how camps are structured and that the camps that include a developmental approach (as opposed to a confrontational approach) are most successful at reducing recidivism.

Regardless of the type of intervention employed, whether traditional juvenile programs or newer adult-type, juveniles involved in arson need to be held accountable. Much already is known about what measures work and how to reach teen firesetters. USFA and the Office of Juvenile Justice and Delinquency Prevention (OJJDP) have documented the program characteristics and models that have proven successful. A 1987 OJJDP/USFA initiative to assess effective programs addressing juvenile firesetting identified seven critical components:

1. **A program management component to** make key decisions, coordinate interagency efforts, and foster interagency support.

2. **A screening and evaluation component to** identify and evaluate children who have been involved in firesetting.

3. **An intervention services component to** provide primary prevention, early intervention, and/ or treatment for juveniles, especially for those who have already set fires or shown an unusual interest in fire.

4. **A referral component to** link the program with the full range of community support agencies that might help identify juvenile firesetters and provide services to them and their families.

5. **A publicity and outreach component to** raise public awareness of the intervention program and encourage early identification of juvenile firesetters.

6. **A monitoring component to** track the program's identification referrals and treatment of juvenile firesetters.

7. **A juvenile justice system component to** establish relationships with juvenile justice agencies that often handle juvenile firesetters.

As part of the USFA OJJDP research, two programs designed to keep difficult children in school and out of trouble were carefully evaluated. One program is called **Communities In Schools (CIS)** and is a network of local, State, and national partnerships working together to bring at-risk youth four basics every child needs: a personal one-on-one relationship with a caring adult who provides support and advocacy; a safe place to learn and grow; a marketable skill to use upon graduation; and a chance to give back to peers and community. There are three essential elements in establishing a local CIS program:

1. a tax-exempt corporation with a board of directors that represents the public and private sectors of the community and that is chaired by a member of the private sector;

2. a management team led by an executive director; and

3. a new education, health, and human services delivery system that repositions or reassigns the community's service resources and focuses them on at-risk youth and their families. Often this results in an alternative program within an existing school.

Outcome studies conducted in South Carolina and Georgia by The Urban Institute demonstrated that high proportions of CIS students remain in school and graduate, and students with the lowest grade point averages raised their averages a full grade point. For further information, contact: Communities In Schools, Inc. located in Alexandria, Virginia (703) 519-8999) *http://www.ncirs.org/ojjdphtmlsafety.html*

The other program is called **SafeFutures: Partnerships To Reduce Youth Violence and Delinquency**. This is a program sponsored by the Office of Juvenile Justice and Delinquency Prevention (OJJDP). OJJDP has provided approximately $1.4 million a year to six communities across the United States (Boston, MA, Contra Costa County, CA, Seattle, WA, St. Louis, MI, Imperial County, CA, and Harlem, MO) for five years to develop a continuum of care responsive to youth and their families at any point along the path toward delinquency. The programs are collaborative efforts which involve local, State, and national agencies. OJJDP is providing technical assistance and training along with such partners as Boys and Girls Clubs of America, Communities In Schools, and the Johns Hopkins University Institute for Policy Studies.

The SafeFutures approach is structured to address offender accountability, apply graduated sanctions, and offer targeted services. The program consists of immediate intervention and sanctions on the first level; and secure confinement in community settings, training, and aftercare on the second.

Follow up and aftercare were cited as critical components for success. The Project Coordinator for SafeFutures at OJJDP can be reached at (202) 307-5914, 1-800 638-8736, NCIRS.

The juvenile court system should aggressively support programs that address juvenile problems through a continuum of services and sanctions that consider youth needs, community safety, and victim reparation.

IMPROVING INFORMATION SHARING ON JUVENILES

Juvenile records confidentiality concerns are frequently cited as a roadblock to effective intervention. Often the Family Education Rights and Privacy Act is cited as a reason why information cannot be shared. Agencies and institutions may be applying an overly restrictive interpretation of this law. In fact, the law was amended in 1994 to allow and facilitate information sharing on juveniles. Educators are permitted to share information with juvenile justice system personnel prior to adjudication, pursuant to State statutes. In **all** circumstances, information can be shared with the consent of a juvenile's parent or guardian. It is critical that agencies serving children and their families share information that enables them to coordinate and provide more effective services. A failure to provide information generally results in fragmentation and duplication of services.

There are other circumstances in which information can be shared among agencies and schools. Information can be provided when a school initiates legal action against a student, when a lawfully issued subpoena is presented, when information about disciplinary action taken against a student is being provided to another school that has a significant interest in the student's behavior, and when a law enforcement record is created by an arm of the educational agency or institution. It is easier to exercise these allowances if the relevant agencies are signatories to a memorandum of agreement covering these points.

ALTERNATIVE PLACEMENTS FOR JUVENILE FIRESETTERS

The treatment for firesetting generally follows the traditional mental health continuum of care which gives priority to the least restrictive environment. Many firesetters can be maintained in the community, often at home, if there is sufficient supervision and responsiveness. Careful assessment is important in determining the proper level of care. A thorough assessment takes into consideration the individual, family, environment, facts about the fire and other fire history, as well as the child's reaction to the fire and sense of accountability.

Sometimes it is determined that the juvenile should be confined to a secure facility, residential treatment center, or hospital. Many programs will not admit a juvenile with a history of firesetting for fear that the child will burn the facility. Interestingly, research indicates that a surprising number of clients in residential facilities actually have a history of firesetting. However, the firesetting was not necessarily identified before placement because the question was never raised. Some studies estimate that upwards of 20 percent of hospitalized juveniles have set a fire. There **are** some alternatives for placing juveniles with a history of firesetting, which have been relatively successful.

Foster Care

Because of the strong correlation between neglect and abuse and firesetting, placing a young person in a safe, supervised family setting can be an effective intervention. Foster care is often more available than institutional placement, and is considerably less expensive.

Research indicates that when firesetting is a cry for help or an effort to bring attention to a serious family situation such as chronic neglect or abuse, removing the stressors stops the firesetting behavior. Foster parents can be trained to work with juvenile firesetters. Intensive foster care programs have been successful in upstate New York and Oregon. Certain foster homes are designated as "intensive" foster homes which qualifies them for more difficult placements, such as older juveniles with firesetting histories. These homes are selected based on the experience of the foster parents, the number and ages of other children placed there, and their willingness to take on higher risk youth. Considerable attention is placed on fire safety practices. Exit drills are practiced regularly, smoke detectors are installed in additional rooms, and safe fire use is emphasized. No fire-related responsibility is given to the foster child initially other than to assist in fire safety precautions. Searches of the foster children's bedrooms, belongings, and person are conducted to be certain that ignition materials are not available. This is agreed to as a prerequisite for admission to the foster home.

"Intensive" foster parents receive training in working with difficult adolescents, which includes communication and problem solving skills, supervision and restraint, behavior management, and fire safety education for prevention and intervention. They also receive considerable support from the social service case workers, including home visits at least a few times a week. Also, the foster children receive a higher level of counseling and support services when appropriate. Parents are included in the treatment plan.

While there is much demonstrated success in these situations, the inherent risk needs to be acknowledged. It is imperative that other children (non-firesetters) placed in this environment be included in the fire safety training and made aware of the potential danger. They should also be taught that if they become aware of fire activity that they need to tell an adult. Keeping people safe in a dangerous situation is not "tattling."

The State of Oregon Juvenile Firesetter Task Force has developed a comprehensive training package for residential treatment personnel. A training videotape is in production. By helping caretakers to better understand the children's behavior and their motivations, the training promotes confidence among the treatment providers. Some facilities stipulate that the juveniles sign "contracts" not to use fire, and use polygraph machines to verify their veracity. These programs also institute firm strictures which include regular searches of rooms and belongings for ignition materials. This practice is crucial if youths leave the facility to attend school or receive services.

The Hillside Children's Center in Rochester, New York, is a comprehensive residential treatment facility providing services to adolescents. The Center was initially reticent about accepting juveniles with a history of firesetting. Subsequent training and experience led to documented success with this at-risk population. Currently, the Center maintains an emergency bed for firesetters who are identified as needing placement by the City of Rochester Fire Department or by the Monroe County Fire Bureau.

Hospitalization

Inpatient facilities are often reluctant to accept adolescents with a history of firesetting. Issues of supervision are often cited, but more often clinicians are concerned that they do not have an effective treatment protocol for "these kids."

Dr. David Kolko at the University of Pittsburgh Medical Center has successfully treated firesetters for several years. The inpatient treatment he uses incorporates intensive individual, group, and family counseling. The counseling uses a cognitive treatment approach which challenges the child's assumptions and rationalizations behind the antisocial behavior--such as burning the school because a teacher disciplined the student.

The treatment at the medical center is skills based. Particular emphasis is placed on providing specific life skills, including interpersonal and problem solving skills. The impact of teaching these skills to delinquent children who were placed in New York State Division for Youth Facilities has been carefully evaluated by Dr. Arnold Goldstein of the University of Syracuse (NY). Rearrest rates were significantly reduced, especially when parents were included in the training.

Irrespective of the seriousness of an incident or the child's motive in starting a fire, education regarding fire should be part of the intervention strategy. Such education should include information about the nature of fire, how rapidly it spreads, and its potential for destructiveness. Discussions about maintaining a fire safe environment, escape plans and practice, and appropriate use of fire have been shown to be effective parts of comprehensive arson intervention programs, at least for younger juveniles.

Similar intervention protocols have been implemented in several cities including Portland, Oregon, Phoenix, Arizona, and Upland, California.

EXAMPLES OF SUCCESSFUL COMMUNITY INTERVENTION PROGRAMS

Lane County Department of Youth Services

The Lane County Department of Youth Services in Oregon operates "CATCH" (Community Alternatives to Commitment Hazards), an intensive probation program focusing on 50 of the worst juvenile cases annually. All juveniles adjudicated for first-and second-degree arson are handled through this well-targeted program which contains most of the effective program elements identified earlier in this report.

CATCH has been formally evaluated and preliminary results show that 93 percent of the youths participating have no subsequent firesetting behavior. Sixty-seven percent have no repeat criminal behavior of any kind. Those who do reoffend are involved in minor incidents such as a curfew violation.

The CATCH program consists of a life skills development curriculum designed for ages 13 to 17 called **Skill Building Curriculum for Juvenile Firesetters.** It is delivered in a group format over 16 sessions by a fire service professional and a youth counselor. The program covers identification of feelings, anger management, empathy training, assertiveness, the confrontation of thinking errors, and a personal fire graph and fire safety. Juveniles are guided toward understanding their firesetting behavior, and are taught skills for coping in positive ways. Parents are also involved in the program and work on parenting skills as well as the skills their children are learning.

The juvenile signs a "contract" which specifies the work to be performed. The contract is reviewed and signed by all involved parties, including the parents. The program includes the skills curriculum as well as three projects which must be completed. The juveniles are required to generate a community impact report, a research project, and a community service project. In the community impact report, the juvenile identifies people or agencies affected by the fire and selects three to five people to interview. The youth asks how the fire affected the victim(s), and completes a report summarizing the interviews. The final report is presented to the court, juvenile counselor, youth services team, or juvenile firesetter network.

For the collage project, the youth is assigned the task of reviewing local newspapers for a specified period of time to collect articles relating to fires. They paste the articles onto poster board and write a report summarizing the headline and description of each article, fire deaths and injuries, dollar loss, cause of the fire, and any other pertinent information about who may have started the fire. These are submitted to the fire department and often are displayed.

The community service project is an opportunity for the youth to learn skills and acquire a sense of giving to the community. Some possible organizations for community service are local parks and recreation departments, food banks, homeless shelters, service clubs, and agencies such as the American Red Cross, Humane Society, or Salvation Army. Community service is also a way of holding the child accountable and offering restitution to the victim. Upon completion of the program the youth explains what has been learned to the overseeing person or team. Following is an example of how the program succeeded with one case.

> A fifteen-year-old male responsible for a $3.5 million dollar fire in his school is a graduate of this program and is a productive member of society today. After intensive intervention, he acquired his high school graduate equivalency diploma and is employed. He has had no subsequent involvement with fire or any other criminal behavior.

Cobb County Juvenile Court Mediation Program

Cobb County, Georgia, Juvenile Court has developed a program designed to "end conflicts with win/win solutions." The program serves Cobb County youths up to 18 years of age, and targets elementary and middle school age children who are first time firesetters with the hope of preventing future offenses. The Juvenile Firesetter Program and the juvenile court refer cases to the Cobb County Juvenile Court Mediation Program. This option is offered before adjudication, and if the program is successfully completed, there is no formal legal record.

The Cobb County Juvenile Court Mediation Program was one of the first court-affiliated intervention programs for youth in the country. The primary goal of the program is to get offending youths to take responsibility for their decisions, be accountable, and modify their destructive behavior. It forces juvenile offenders to come face-to-face with their victims and to negotiate acceptable restitution. The intervention program also includes an education component which teaches fire prevention, decision making, and consequences for behavior.

The Cobb County Juvenile Court Mediation Program uses both teen and adult mediators. Staff mediators complete rigorous training provided by the University of Georgia. The program addresses both delinquent and status offenses. In Georgia, delinquent offenses include youth who are ungovernable, violate curfew, and run away from home. Status offenses may include firesetting, shoplifting, simple assault and battery, and damage to property.

The mediation process includes the offending youth, at least one parent or legal guardian, and the victim/complainant. The process allows the victims to be heard and provides restitution. Acceptable forms of restitution include monetary compensation, yard/house work, services rendered to a local merchant, or community service.

The mediation agreements are binding agreements. In addition, new problem solving skills are modeled and learned. The process is less costly than litigation and it benefits the parties as well as the community.

The Cobb County program is an excellent example of the concept referred to as "balanced and restorative justice." The program reinforces that when a crime is committed the offender is responsible for restoring the victim to a pre-offense condition.

Phoenix Fire Department Juvenile Firesetter Task Force

The Phoenix Fire Department maintains one of the longest running and most successful juvenile firesetter programs in the United States. The program is staffed by full-time fire investigators and fire safety educators so that cases are handled expeditiously and comprehensively. What is particularly unique about this program is the well-established community network that supports the program on an on-going basis.

A task force made up of representatives from community agencies meets on a regular basis to review individual cases. The group also monitors gaps in services to best meet the needs of the youth who are referred to the firesetter program. Juvenile court officials, including a hearing officer and a prosecutor, are active members of the task force. Also represented are educators, mental health professionals, law enforcement officers, fire investigators, and child welfare workers. Such broad involvement ensures that all levels of service--from prevention education to post treatment placement and a return to the community--are delivered.

In partnership with the task force, St. Luke's Hospital has helped establish a range of community services and coordinates closely with the Phoenix Fire Department and the juvenile court system. Grants have been obtained that fund comprehensive assessment and intervention services at no cost to the juvenile or the parents. A referral network has been established to provide support to treatment providers, including residential treatment centers. The task force also is helpful in educating other providers about juvenile firesetting and breaking down barriers to services and placement.

CONCLUSIONS

Older juveniles involved in intentional firesetting are a serious problem in the United States. This form of arson accounts for thousands of fires, hundreds of deaths, thousands of injuries, and millions of dollars in direct property loss every year. **Nationally, juveniles now account for the majority of arson fires.**

Fires set by older juveniles closely follow the patterns reported by USFA and NFPA.

Occupied structures are significantly represented and include dwellings and schools. Vacant buildings are at increased risk, as are outdoor areas which include dumpsters, playgrounds, and parks.

Case studies suggest that different motivations may influence different targets. Gang-related or vandalism fires often target abandoned buildings and dwellings. Older juveniles who are troubled and responding to family crises often set fires in their own homes or schools. Juvenile firesetting dynamics should be studied more thoroughly, the results of which could lead to more effective interventions through the identification of specific profiles.

Older juvenile firesetters are often alienated, angry, and adept at acting out through various forms of destruction. The increase in violent juvenile crime, including firesetting, has led to changes in State and Federal laws which now allow juveniles, in some circumstances, to be prosecuted as adults. Some studies of juveniles transferred to adult courts indicate that these teens commit more crimes upon their release. Juvenile courts should consider programs that address juvenile problems through a comprehensive continuum of services and sanctions that take into account community safety, victim reparation, and youth needs. Programs that have incorporated restitution and community service along with skill building and individual and family support have been documented to reduce firesetting.

Perhaps most important, juvenile courts along with the fire service and other community agencies need to prioritize firesetting cases. The community's intervention should be swift and decisive with consistent, predictable consequences. Juvenile firesetter intervention programs need to be supported or enhanced so that detection and assessment takes place quickly. Early detection and intervention improves the likelihood of preventing future firesetting.

BIBLIOGRAPHY

Adolescent Firesetters: An Intervention. (1997). Hot Issues, Volume 7, No. 2. Salem, Oregon: Juvenile Firesetter Intervention Program.

Bills, J., Cole, R, Crandall, R., & Schwartzman, P. (1990). Fireproof Children Handbook. Rochester, NY: National Fire Service Support Systems, Inc.

Bishop, D., Frazier, C., Lanza-Kaduce, L., & Winner, L. (1996). The Transfer of Juveniles to Criminal Court: Does It Make a Difference? Crime & Delinquency, Volume 42, No. 2, 171-191.

Campbell, C., & Elliott, E. (1997). Skills Curriculum for Intervening with Firesetters (Ages 13-18). Oregon: Stop-Fire.

Cole, R., & Schwartzman, P. (1992) Rochester City School District Long-term Suspension Project: Preliminary Evaluation Report. Submitted to the Superintendent of Schools.

Cole, R., Grolnick, W., Laurenitis, L., McAndrews, M., Matkowski, K., & Schwartzman, P. (1986). Children and Fire: Rochester Fire-Related Youth Project progress report. Rochester, NY.

Cole, R, Grolnick, W., & Schwartzman, P. (1993). Fire Setting. In Ammerman, R, Last, C., & Hersen, M. (Ed), Handbook of Prescriptive Treatments for Children and Adolescents. Boston: Allyn & Bacon.

Cole, R., Laurenitis, L., McAndrews, M., McKeever, J., & Schwartzman, P., (1983). Juvenile firesetter intervention: Report of the Rochester Fire-Related Youth Program Project. Rochester, NY.

Fagan, Jeffrey. (1997). The Comparative Advantage of Juvenile Versus Criminal Court Sanctions On Recidivism Among Adolescent Felony Offenders. A Report to the National Institute of Justice. Washington D.C.: U.S. Department of Justice.

Gang-related Fires in Phoenix. (1997). Hot Issues, Volume 7, No. 3. Salem, Oregon: Juvenile Firesetter Intervention Program.

Goldstein, Arnold, P. (1991). Delinquent Gangs: A Psychological Perspective. Champaign, Illinois: Research Press.

Grolnick, W., Cole, R., Laurenitis, L., & Schwartzman, P. (1990). Playing with fire: A developmental assessment of children's fire understanding and experience. Journal of Clinical Child Psychology, 19, 128-135.

Hall, John, R., Jr. (1995). U.S. Arson Trends and Patterns--1994. Quincy, MA: National Fire Protection Association.

Hall, John, R, Jr. (1995). Children Playing With Fire: U.S. Experience, 1980 - 1993. Quincy, MA: National Fire Protection Association.

Kolko, D.J., Dorsett, P.G., & Milan, M.A. (1981). A Total Assessment approach to the evaluation of social skills training: The effectiveness of an anger control program for adolescent psychiatric patients. Behavioral Assessment, 3, 383-402.

Kolko, D.J., & Kazdin, A.E. (1986). A conceptualization of firesetting in children and adolescents. Journal of Abnormal Child Psychology, 4, 49-61.

U.S. Department of Justice. (1996). What Works: Effective Delinquency Prevention and Treatment Programs. Juvenile Justice Fact Sheet #20.

U.S. Fire Administration. (1997). Children and Fire: The Experience of Children and Fire in the United States. Washington, DC: National Fire Data Center.

U.S. Department of Justice. (1996). The Gould-Wysinger Awards: A Tradition of Excellence. Juvenile Justice Fact Sheet #44.

U.S. Department of Justice. (1996). Information Sharing and the Family Educational Rights and Privacy Act. Juvenile Justice Fact Sheet #39.

U.S. Department of Justice. (1997). Adolescent Motherhood: Implications for the Juvenile Justice System. Juvenile Justice Fact Sheet #50.

U.S. Department of Justice. (1995). Boys & Girls (B&G) Clubs of America. Bureau of Justice Assistance Fact Sheet.

U.S. Department of Justice. (1997). Detention and Delinquency Cases, 1985-1994. Office of Juvenile Justice and Delinquency Prevention Fact Sheet #56.

U.S. Department of Justice. (1994). Researchers Evaluate Eight Shock Incarceration Programs. National Institute of Justice Update.

U.S. Department of Justice. (1996). SafeFutures: Partnerships To Reduce Youth Violence and Delinquency. Juvenile Justice Fact Sheet #38.

U.S. Department of Justice. (1997). Juvenile Firesetting and Arson. Juvenile Justice Fact Sheet #51.

U.S. Department of Justice. (1996). Balance and Restorative Justice Project. Juvenile Justice Fact Sheet #42.

U.S. Department of Justice. (1996). Serious Habitual Offender Comprehensive Action Program. Juvenile Justice Fact Sheet #35.

www.ingramcontent.com/pod-product-compliance
Lightning Source LLC
Chambersburg PA
CBHW081246170526
45165CB00009B/3218